Bruno Duarte
Lucas Lima Pereira
Arthur Santos Lopes

Efficiency of an Indirect Expansion Cooling Simulator

Bruno Duarte
Lucas Lima Pereira
Arthur Santos Lopes

Efficiency of an Indirect Expansion Cooling Simulator

Performance study of an indirect expansion cooling simulator using an Arduino microcontroller

ScienciaScripts

Imprint

Cover image: www.ingimage.com

This book is a translation from the original published under ISBN 978-620-6-76074-0.

Publisher:
Sciencia Scripts
is a trademark of
Dodo Books Indian Ocean Ltd. and OmniScriptum S.R.L publishing group

120 High Road, East Finchley, London, N2 9ED, United Kingdom
Str. Armeneasca 28/1, office 1, Chisinau MD-2012, Republic of Moldova, Europe
Printed at: see last page
ISBN: 978-620-8-12718-3

SUMMARY

In this final paper, a superficial analysis of the performance of a large refrigeration simulator was carried out for the Mechanical Engineering course at UEMA. Initially, a historical and thermodynamic approach is taken to refrigeration cycles, refrigerants, the equations and diagrams characteristic of the subject, as well as describing the operation of the various mechanical and electrical components and their importance for the proper functioning of the installation. The data was collected using an open-source electronic platform based on software and hardware (Arduino), with an interface using the EXCEL programming language, VBA, in the laboratory. Vacuuming and refrigerant gas recharging procedures were carried out on the installation to ensure better machine performance. At the end of the study, we found that there were possible refrigerant leaks, which resulted in poor machine performance, with little capacity to absorb heat in the evaporator.

Key words: Performance, Temperature, Cooling, microcontroller.

ABSTRACT

With this work of completion of course, a superficial analysis of the performance of a large refrigeration simulator of the mechanical engineering course of the UEMA was carried out. Initially, a historical and thermodynamic approach to cooling cycles, refrigerant fluids, equations and diagrams characteristic of the theme is described, as well as describing the operation of the various mechanical and electrical components and their importance for the proper functioning of the installation. Data collection was performed with the help of an open source electronic software platform based on software and hardware (Arduino), with an interface with the programming language of EXCEL, VBA in the laboratory. For this purpose, the vacuum and refill procedures of the refrigerant gas were made in the installation in order to ensure a better performance of the machine. In this study, we verified at the end of the study that there are possible leaks of refrigerant fluid, which resulted in low machine performance, with little capacity to absorb heat in the evaporator.

Keywords: Performance, Temperature, Refrigeration, microcontroller

LIST OF ABBREVIATIONS, ACRONYMS AND SYMBOLS

QHot source heat QCold source heat Wliq Net work

TRTon of refrigeration UR Relative humidity

lbmlibra pasta

°CDegrees Celsius

°FDegrees Fahrenheit

inPoleged

HgMercury

PsiPounds per square inch absolute - pounds per square inch absolute. hhoras

COP-B Coefficient of performance of the heat pump

COP-R Coefficient of performance of the refrigerator kjKilo-Joule

minminute

kj/minKilo-Joule per minute BTU British Thermal Unit m² square meter

cubic meter

kwKilo-watt

kgKilo-gram

kPaKilo-Pascal

Δ tTemperature change

T-STemperature Entropy

₵ *c₵_p* Specific heat at constant pressure PUC Pontifical Catholic University

SUMMARY

1INTRODUCTION

The concern with preserving food and being in a mild climate has always been present in human life since primitive times. For many years, natural ice was the main known cooling element and was stored in various ways, whether in caves, holes or trenches, according to the chronological order of evolution (SERAFIM MATOS, 2017).

The first refrigerant used was water, until the refrigerants we know today were developed. The first refrigeration cycle to be put into operation was by Jacob Perkins in 1834. However, this work was not widely accepted by the scientific community of the time, and the man responsible for developing mechanical compression in real equipment was the Scotsman James Harrison, with his patents in 1856 and 1857 under the respective patents 747 and 2362. (SERAFIM MATOS, 2017)

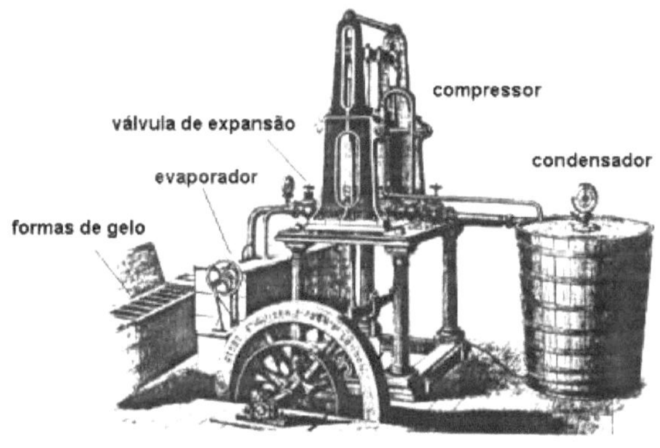

Figure 1.1 James Harrison's mechanical vapor compression refrigeration machine

Source: CORRÊA (2017)

With technological developments and the discovery of electricity, the first automatic refrigerator powered by electricity appeared in 1918, manufactured on a small scale by the Kevinator Company in the United States. It wasn't until 1928, with advances in the study of refrigerants and the use of environmentally friendly substances, that refrigeration entered an industrial scale (SERAFIM MATOS, 2017).

Since then, there have been various ramifications and applications in the field of refrigeration and air conditioning. Today, various sectors of the economy use cold in their processes, such as the chemical, oil and petrochemical, food, manufacturing, air conditioning and heating industries, such as air conditioners and heat pumps, among others. It is a vast market for engineers, who can work in areas such as research, development and application of equipment with a strong tendency towards air conditioning and industrial refrigeration (Ar condicionado e refrigeração, 2017).

One of the most important factors that allow refrigeration to take place is the ability of a refrigerant to absorb heat when evaporated, and this evaporation can be achieved by mechanical processes, for example. There are several advantages to using refrigerants, such as the ability to be reused in the cycle, easy control of phase changes and durability. There are various applications for refrigeration, such as domestic refrigeration, using refrigerators and freezers; commercial refrigeration, found in stores, supermarkets and hotels; industrial refrigeration, found in factories, refineries, food processing plants and requiring more technical labor; maritime and transport refrigeration, with ships in general and buses, cars, trucks and railway wagons; cryogenic refrigeration, for very low temperatures and air conditioning. (SERAFIM MATOS, 2017)

The vapor compression system is one of the most commercially and industrially applied and originates from the reverse Carnot cycle, in which heat is removed from a cold source to a hot source using mechanical work to do so (Cengel, 2013). There are direct and indirect expansion refrigeration cycles with their own particularities. In the first case, the air to be cooled (hot source) is forced by fans to give off heat directly to the refrigerant (cold source) which is circulating in an evaporator coil. In the second case, as in chillers, the subject of this paper, the air to be cooled is expanded by water, which is an intermediate element. The closed gas cycle cools the water which is stored in a container (thermoaccumulation) and then pumped to the Fancoil where it cools the room (Refrigeration Systems: Direct Expansion vs. Indirect Expansion, 2016).

With the water already cold, it is then distributed through pumps, pipes and valves. By means of cooling coils and heat exchangers, the thermal load of the air is transferred to the water and returned to the chiller. There is a wide variety of chillers on the market, the most common of which are absorption, centrifugal, helical rotation and scroll and some alternative chillers (Chiller System Design and Control, 2009).

These types of air-conditioning systems, which use water to cool the air, have several advantages over air-operated systems, such as longer service life and operating efficiency, and less physical space. However, proper design and maintenance are necessary for good results (III, Herbert W. Stanford).

2 OBJECTIVES

2.1 GENERAL OBJECTIVE

The general aim of this work is to analyze the performance of a direct expansion refrigeration simulator for a practical class in the refrigeration laboratory of the Mechanical Engineering course at the State University of Maranhão-UEMA, using temperature and pressure data collected from the machine.

2.2 SPECIFIC OBJECTIVES

- Evacuate high and low pressure pipes
- Refill with suitable refrigerant gas.
- Monitor components for possible leaks, excessive noise and other abnormalities.

3 THEORETICAL BASIS

3.1 REFRIGERATION

Refrigeration is the process of transferring heat from a lower temperature region to a higher temperature region, as seen in Figure 3.1. For this to be possible, it is necessary to use so-called chillers or heat pumps that operate in cycles and use working fluids called refrigerants. The difference between the two is that the purpose of chillers is to remove heat from a refrigerated medium and heat pumps is to supply heat to a heated environment. (Cengel, 2013)

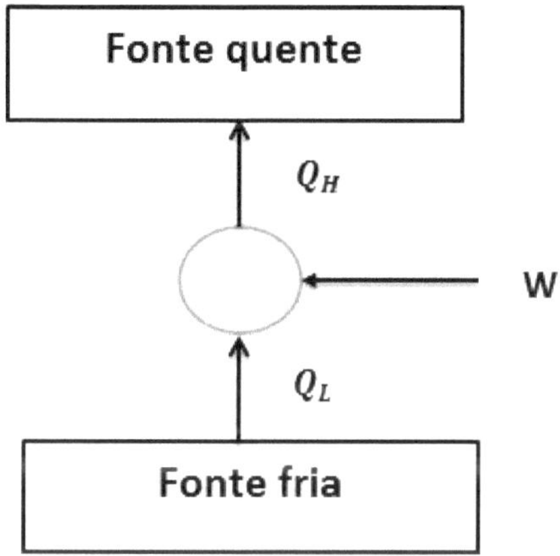

Figure 3.1: Refrigeration process Source: Author

The figure above shows the schematic of a refrigeration process, where Q_L is the magnitude of the heat removed from the refrigerated space at temperature tc and Q_H is the magnitude of the heat rejected at temperature th and W is the net work input into the refrigerator (Cengel, 2013).

3.2 COEFFICIENT OF PERFORMANCE (COP)

The performance of chillers and heat pumps is defined in terms of the coefficient of performance (COP), defined as follows: Equation 3.1 and 3.2.

COPR = $(Resultado\ desejado)/(Entrada\ necessária)$
= $(Efeito\ de\ resfriamento)/(Entrada\ de\ trabalho)$ = $\frac{Q_L}{Wliq_R}$

(3.1)

COPB = $(Resultado\ desejado)/(Entrada\ necessária)$

= $(Efeito\ de\ resfriamento)/(Entrada\ de\ trabalho)$ = $\frac{Q_H}{Wliq_B}$

(3-2)

These expressions can be expressed in rate form by substituting Q-H and $Wliq$

$\dot{Q}_L$, $\dot{Q}_H$, $\dot{w}liq$,

respectively. To measure the cooling capacity of a system, i.e. the rate at which heat is removed from a refrigerated space, it is usually referred to by the ton of refrigeration (TR). This TR is the cooling capacity that can freeze 1 ton (2000lbm) of liquid water at 0°C (32°F) and turn it into ice at 0°C in 24 hours. One ton of refrigeration is equivalent to 211 kj/min or 200 Btu /min and the cooling load of a typical $200m$^2 home is in the range of 3ton (10 kw). (Cengel, 2013).

However, more commonly found in air conditioners is the BTU (British Thermal Unit), a British heat unit. In other words, the BTU is the amount of heat needed to raise the temperature of 0.45 kg of water by 0.56 °C and is equivalent to 1.055 Joules and in relation to TR, we have 1 TR=12000 BTU.

3.3 VAPOR COMPRESSION REFRIGERATION CYCLE

3.3.1 Carnot cycle reverse

The Carnot cycle is the one that has the maximum thermal efficiency between certain temperature limits, is fully reversible and serves as a standard of comparison for real power cycles. As the cycle is reversible, the four processes that comprise it can also be covered by reversing the direction of the work and heat interactions in the T-s diagram (Çengel, 2013). In order to remove heat from a cold source to a hot source, external work is required. The processes that occur in the cycle are:

- ✓ **1-2 isentropic compression;**
- ✓ **2-3 isothermal heat rejection;**
- ✓ **3-4 isentropic expansion;**
- ✓ **4-1 isothermal heat absorption**

The extraction of heat from the low-temperature source in process 4-1 is the purpose of the cycle. All the other processes in the cycle take place to transfer the energy taken from the low-temperature source to a convenient high-temperature reservoir (Stoecker & Jerold W. Jones, 2017).

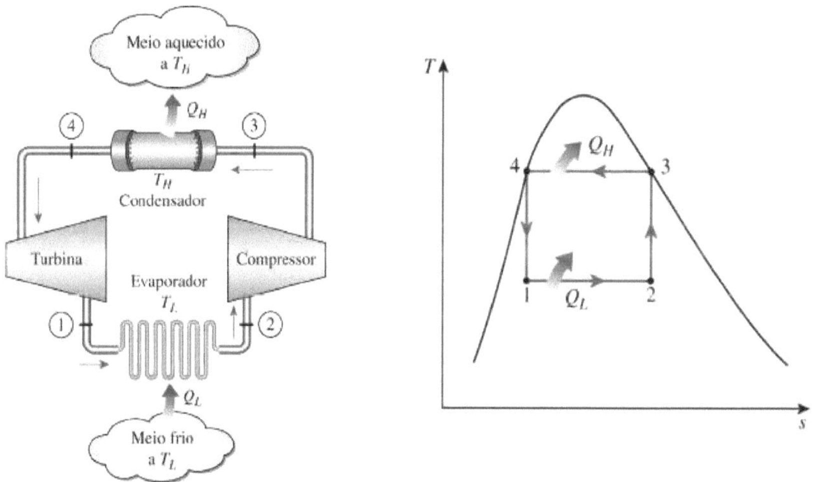

Figure 1.2 - Diagram and T-s cycle of the reverse Carnot cooler

Source: (Cengel, 2013)

$$COP\_R = 1/(T\_H/T\_L - 1) \quad (3\text{-}4)$$

The performance coefficients of chillers and heat pumps are expressed by:

$$COP_B = \frac{1}{\frac{T_L}{T_H} - 1} \quad (3\text{-}5)$$

We can see that the COP's increase the smaller the temperature difference between the cold and hot sources. However, in practice the Carnot refrigeration cycle is not easy to achieve since the 2-3 compression process involves a mixture of liquid and vapor, which would require a compressor that operates with both phases and the 4-1 expansion of refrigerant with a high moisture content in the turbine. The costs and difficulties of operation make it unfeasible, but it serves as a model for comparing real cycles (Çengel, 2013).

3.3.2 Ideal compression refrigeration cycle vapor

A refrigeration vapour compression cycle is a thermodynamic process in which a refrigerant fluid is vaporized, compressed and condensed through a continuous cycle to allow a given space to be cooled. (SERAFIM MATOS, 2017)
Replacing the turbine with a capillary tube or expansion valve resulted in the ideal vapor compression refrigeration cycle. The processes in this cycle are:

✓ 1-2 Reversible adiabatic compression from the saturated vapor state to the condensing pressure;

✓ 2-3 Reversible rejection of heat at constant pressure, lowering the temperature of the refrigerant initially and condensing it later;

✓ 3-4 Irreversible expansion at constant enthalpy from the saturated liquid state to the evaporation pressure;

✓ 4-1 Heat gain at constant pressure, producing evaporation of the refrigerant to the state of saturated vapor

This cycle is shown in the figure below:

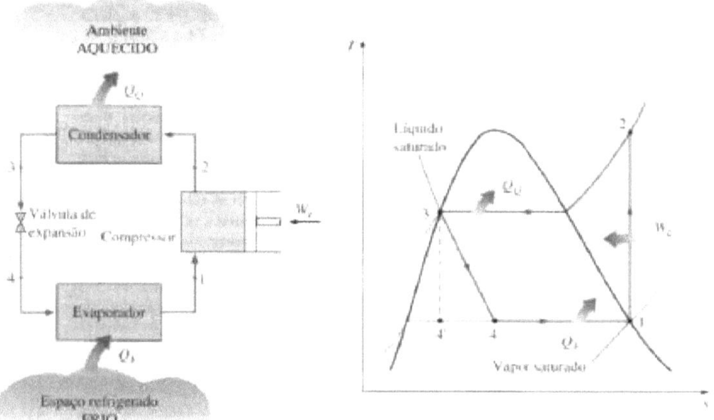

Figure 2: Ideal vapor compression cycle
Source: Adapted from (Cengel, 2013)

In the evaporator, the refrigerant gas receives heat from the refrigerated space 4-1 and changes phase from liquid to vapor when it reaches the evaporation temperature. This temperature must be sufficient for the evaporation temperature to be lower than that of the fluid to be cooled. To maintain the evaporation pressure, the compressor draws in vapor at the same rate as it is produced in the evaporator. In the 1-2 compressor, the vapor pressure is raised until the saturation temperature corresponding to the discharge pressure is higher than that of the cooling fluid; this fluid is usually atmospheric air or water (CORRÊA, 2017). The cooling fluid removes heat from the refrigerant, which changes phase from vapor to liquid at condensing pressure 2-3. The saturated liquid leaves the condenser, passes through the expansion device and has its pressure reduced 3-4. In the expansion process (pressure reduction) part of the liquid evaporates, so the refrigerant leaving the expansion device towards the evaporator is a liquid-vapor mixture at the evaporation pressure. (CORRÊA, Apostila de refrigeração e climatização, 2017)

The vapor fraction in the mixture (called the gas flash) does not produce useful cooling in the evaporator, but it has already served its purpose of reducing the temperature of the rest of the liquid.

This fraction of vapor in the mixture is important practical data; although it is a small fraction of the total mass flow (approximately 15%), in terms of volumetric flow it can reach relatively significant values (98% for ammonia). (CORRÊA, Apostila de refrigeração e climatização, 2017)

In the figure above, if the expansion device were replaced by a turbine, the refrigerant would enter the evaporator in state 4' which would increase the power of the refrigerant effect, i.e. absorbing a greater amount of heat. However, the benefits do not justify the additional costs and complexity (Çengel, 2013).

In the ideal cycle, the four components that operate in the steam refrigeration cycle can be approximated as devices that flow in a permanent regime. In general, variations in kinetic and potential energy are disregarded.

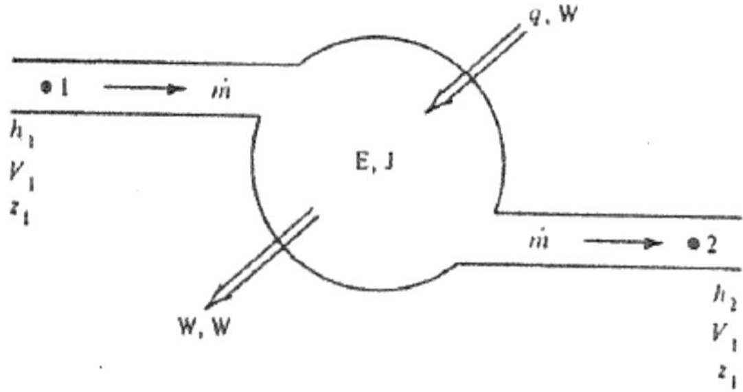

Figure 3: Energy balance in a control volume with constant flow rate
Source: (Stoecker & Jerold W. Jones, 2017)

When calculating the energy balance for a flow in permanent regime, using the energy conservation equation or first law of thermodynamics for a control volume, we have:

$$dE/dT = (q)^{\cdot} - w^{\cdot} + m^{\cdot}(h_1 + (V_1^2)/2 + gz_1) - m^{\cdot}(h_2 + (V_2^2)/2 + gz_2) \qquad (3\text{-}6)$$

Where:

$m^{\cdot}$= flow rate, kg/s

h= enthalpy, j/kg V= velocity, m/s Z= elevation, m

g= acceleration due to gravity= 9.8 m/s²

$q^{\cdot}$= energy transfer rate in the form of heat, ww $^{\cdot}$= energy transfer rate in the form of work, w

In most steady-state processes, dE/dT =0. In this case, the energy equation can be written as:

$$m^{\cdot}(h\_1 + (V\_1\text{^}2)/2 + gz\_1) + (q)^{\cdot} = m^{\cdot}(h\_2 + (V\_2\text{^}2)/2 + gz\_2) + \mathrm{w} \qquad (3\text{-}7)$$

This form of the energy equation makes it possible to calculate the energy interactions involving the components of the refrigeration cycle.

3.3.2.1 Heating and cooling

The evaporator is the component in which the refrigerant, when it passes through the expansion valve, enters at a low pressure and temperature, since in order to absorb heat from the refrigerated space, the refrigerant must be at a lower temperature than the environment and there must be control over the temperature difference between the evaporator and the cooled environment. When the refrigerant leaves the evaporator, it is in the form of steam (SERAFIM MATOS, 2017).

In the condenser, heat is rejected to the external environment. In this component, steam enters at a high temperature and pressure and condenses by exchanging heat with air or water (SERAFIM MATOS, 2017).

By making a control volume in the evaporator or condenser, and disregarding the variations of kinetic energies $v\text{^}2/2$ and potentials gz and no work is done in the process $\mathrm{w} = 0$, we can calculate the heat transfer rates in the evaporator and condenser and equation 2-7 reduces to the following cases: (Stoecker & Jerold W. Jones, 2017).

1) For the evaporator

$$_L Q^{\cdot} = m^{\cdot}\ (h_1 - h\ )_4 \qquad 3\text{-}8)$$

2) For the condenser

$$_H Q^{\cdot} = m^{\cdot}\ (h_2 - h\ )_3 \qquad (3\text{-}9)$$

3.3.2.2 Isentropic compression

Removing heat from a cold source to a hot one requires the use of mechanical work, which is carried out by means of a compressor. This is an electromechanical device that is used to develop and maintain the flow of fluid through the refrigeration system. The cold vapor flows from the evaporator to the compressor where the pressure and temperature are raised making the fluid highly superheated at the outlet of the equipment, causing the process to raise the saturation temperature of the vapor to a level that allows the vapor to condense at normal ambient temperatures. The refrigerant flows through the remaining components in the system and returns to the evaporator where the cycle is repeated (SERAFIM MATOS, 2017).

In the compressor working at steady state, it is considered adiabatic q=0, since heat losses to the external environment are negligible in view of the net energy flows of the system and the variations in kinetic energy and potentials are negligible, making equation 2-7 a:

$$\dot{w} = \dot{m}\,(h_2 - h_1) \qquad (3\text{-}10)$$

In other words, the power required for compression is equal to the product of the flow rate and the enthalpy change. (Air conditioning and refrigeration, 2017)

3.3.2.3 Isentropic expansion

The throttling process is adiabatic with □ = 0, the kinetic and potential energy variations are negligible and there is no work $w = 0$ which reduces equation 2-7 to the following form

$$h_3 = h_4 \qquad (3\text{-}11)$$

With the correct installation and insulation of the valve, there is no heat exchange with the outside, part of the refrigerant that vaporizes absorbing heat from the rest of the liquid mass. Thus, at the outlet of the device we have a mixture of liquid and vapor at a lower temperature and pressure than at the inlet to the metering device (p4<p3 and t4<t3), due to the throttling that occurs in the device. (SERAFIM MATOS, 2017)

In the ideal vapor compression refrigeration cycle, the coefficient of performance can be calculated for chillers and heat pumps.

COP_R=*q_L*/(W_*Liq*, *ent*)=(*h*_1 - *h*_4)/(*h*_2 - *h*_1 (3-12)

□□□_□=□_□/(W_*Liq*, *ent*)=(*h*_2 - *h*_3)/(*h*_2 - *h*_1) (3-13)

A rule of thumb is that the COP improves by 2 to 4% for each °C increase in evaporation temperature or each °C decrease in condensation temperature (Cengel, 2013).

3.3.3 The real compression refrigeration cycle from vapor

For the real cycle, there are several points of energy dissipation, i.e. irreversibilities that somehow cause inefficiency in the cycle as a whole. The main differences between these cycles lie in the pressure losses in the evaporator and condenser, the subcooling of the liquid leaving the condenser and the overheating of the vapor at the compressor suction. Due to these load loss effects, which are not relevant in the ideal cycle, in the real cycle, due to friction, there is a load loss resulting in greater compression work between states 1 and 2 (Air conditioning and refrigeration, 2017).

With subcooling at the condenser outlet, there is a greater guarantee that the liquid refrigerant will go to the expansion valve. On the other hand, superheating the vapor leaving the evaporator prevents liquid droplets from entering the compressor. Another relevant point when comparing the cycles is that the actual compression is not isentropic, resulting in inefficiency due to friction and other losses. (Air conditioning and refrigeration, 2017)

The result of the superheat, which in other words is the heat gain in the connection line and the pressure drops in the evaporator and connection line, is an increase in the specific volume, which also increases energy consumption, since the work with flow in permanent regime is proportional to the specific volume. (Cengel, 2013)

Thus, when it comes to overheating, this can increase or decrease the COP depending on the refrigerant used (CORRÊA, Apostila de refrigeração e climatização, 2017).

Figure 6 and 7 below show the pressure-enthalpy diagrams for subcooling and subheating.

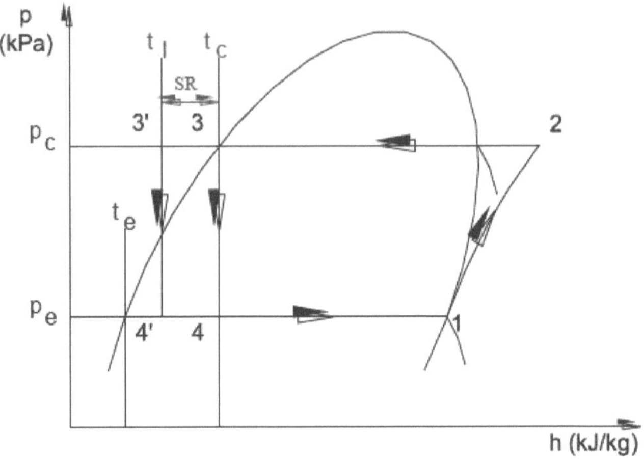

Figure 4: Pressure-enthalpy diagram with subcooling
Source: (SERAFIM MATOS, 2017)

Where:

Pc=condensation pressure, kPa Pe=vaporization pressure, kPa tc=condensation temperature, °C tl=liquid line temperature °C

According to Serafim Matos (2017) we can see from the diagram above that there is an increase in the refrigerant effect due to the greater heat absorption area for the same mass compression rate with subcooling. The design and sizing of the subcooling section In addition, when sizing the liquid line, subcooling can be used to compensate for pressure drops due to friction and other dissipative sources For the cycle with superheating, we have the following graphical configuration:

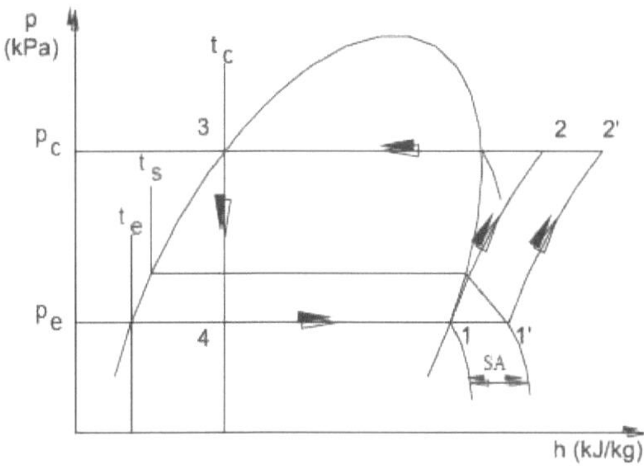

Figure 5: Pressure-enthalpy diagram with overheating
Source (SERAFIM MATOS, 2017)

What usually happens is that after the liquid refrigerant has completely vaporized in the evaporator, the cold saturated vapor will generally continue to absorb heat and, in this way, the vapor reaches the compressor in the form of superheated vapor (SERAFIM MATOS, 2017). The ideal and real cycles can be seen in a comparative pressure-enthalpy diagram as shown in Figure 8.

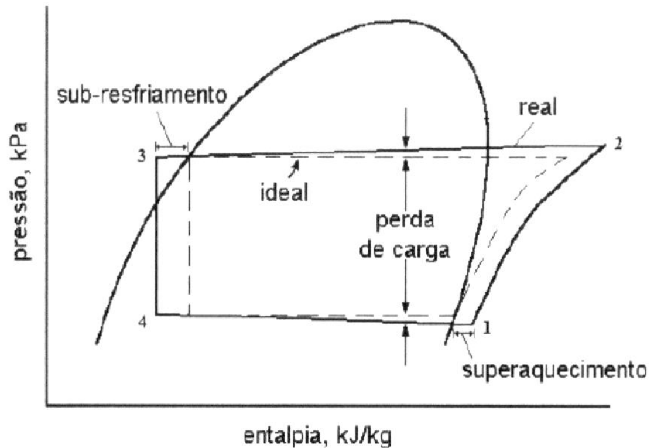

Figure 6: P× h diagram showing the ideal cycle (dashed) and the real one.
Source: (SERAFIM MATOS, 2017)

.

3.4 REFRIGERANTS

A refrigerant is any body or substance that acts as a cooling agent (thermal agent) by absorbing heat from another body or substance. The refrigerant is the working fluid of the cycle in a mechanical vapor compression that alternately vaporizes and condenses when it absorbs and gives off heat, respectively. (A.S. de CAMPOS & A. de SOUZA JR,2009,p.33)

When used in steam compression systems they are called primary fluids and when a liquid is used intermediately to transport thermal energy at a low temperature from one place to another, as in the case of chilled water plants, they are called secondary fluids. Secondary coolants are those known as antifreezes and brines (A. S. de CAMPOS & A. de SOUZA JR, 2009).

3.4.1 Halogenated hydrocarbons

These are hydrocarbons that contain one or more of the following halogens: Cl, F, Br. Hydrogen may or may not appear. The numerical designations of these hydrocarbons are:
1st digit, no. of C-1 atoms (if null, omit) 2nd digit, no. of H+1 atoms
3rd digit, no. of F atoms For example:

CCl3F - Trichloromonofluormethane, R-11 (CFC-11) CHClF2 - Monochlorodifluormethane, R-22 (HCFC-22) CHF2CHF2 - Tetrafluorethane, R-134 (HFC-134) CF3CH2F - Tetrafluorethane, R-134a (HFC-134a)
CBrF3 - Bromotrifluormethane, R-13B1 (B1 indicates no. of Br atoms)
Isomers are distinguished by a criterion of symmetry, based on the atomic masses of the atoms attached to each carbon atom. (SERAFIM MATOS, 2017)

3.4.2 Chlorine-fluorocarbon (CFC)

They are hydrocarbon derivatives obtained by replacing hydrogen atoms with chlorine and fluorine atoms, such as CFC-11, CFC-12, CFC-113, CFC-114 and CFC-115. The lifespan of CFCs in the atmosphere varies from 60 to 540 years, causing ozone depletion. Most CFCs have an ozone depletion potential index. (A.

S. de CAMPOS & A. de SOUZA JR, 2009).

3.4.3 Hydrofluorocarbon (HFC)

This type of soft drink is made up of carbon, hydrogen and fluorine atoms. They are less harmful to the environment as they do not cause ozone depletion because they do not contain chlorine atoms in their composition. These types of fluids include: HFC-134a, HFC-125 and HFC-143a. There is a tendency for HFC fluids to become the most widely used in air conditioning systems in the coming decades (A. S. de CAMPOS & A. de SOUZA JR, 2009).

3.4.4 Security

According to ANSI/ASHRAE standard 34-1992, there is a parameter for measuring the toxicity of refrigerants in Classes A or B. For Class A, they are the least aggressive or the least toxic and least dangerous, i.e. they are not toxic when their concentration is less than or equal to 400 ppm (parts per million) by mass. Class B refrigerants, on the other hand, are the most dangerous and show evidence of toxicity at concentrations of 400 ppm or less.

In practice, there are several examples of Class A refrigerants, such as: HCFC-22, HFC-134a, CFC-11, CFC-12 and R-718 (water), while Class B refrigerants include HCFC-123 and R-717 (ammonia). R-717 (ammonia) mixed with air in concentrations of between 16 and 25% by volume is highly explosive, although it falls into the low flammability group (A. S. de CAMPOS & A. de SOUZA JR, 2009).

A degradation index was created to compare the relative influence on the ozone layer of the various refrigerants, called ODP (Ozone Depletion Potential), which relates the ozone depletion rate of 1 kg of any refrigerant to that obtained for CFC-11. CFC-11's ODP was assigned a value of 1 (one). The GWP (Global Warming Potential) index was also created, used to compare the effects of CFCs, HCFCs and HFCs on global warming based on the effects of CFC-11. Table 5 shows the ODP and GWP values for each of the refrigerants (A.S. de CAMPOS & A. de SOUZA JR, 2009, p.36).

Table 1 below shows some ODP and GWP values for halogenated refrigerants

Table 1: Atmospheric residence time for some refrigerants, ODP (ozone depleting potential) and GWP (global warming potential)

Refrigerante	Tempo de vida (anos)	ODP	GWP
CFC-11	60	1,00	1,00
CFC-12	120	1,00	3,00
CFC-113	90	0,80	1,40
CFC-114	200	0,70	3,90
CFC-115	400	0,40	7,50
HCFC-22	15	0,05	0,34
HCFC-123	2	0,02	0,02
HCFC-124	7	0,02	0,10
HCFC-141b	8	0,10	0,09
HCFC-142b	19	0,06	0,36
HFC-125	28	0,00	0,58
HFC-134a	16	0,00	0,26
HFC-143a	41	0,00	0,74
HFC-152a	2	0,00	0,03

Source: (SERAFIM MATOS, 2017)

3.5 INDIRECT EXPANSION

In indirect expansion refrigeration systems, the chiller plant is located in a machine room. The chilled water is produced by a direct gas expansion system and pumped through insulated pipes and distributed to the thermal load absorption points, the fan-coils. These intake openings are carefully designed to take wind pressure and the chimney effect into account (CORRÊA, Apostila de refrigeração e climatização).

The most common fluids used in chiller systems are water or brine, depending on the application. The intermediate agent, heated by absorbing heat from the refrigerated space or the product, returns to the chiller to be cooled and recirculated (SERAFIM MATOS, 2017). Figure 9 below shows a schematic of a chilled water system.

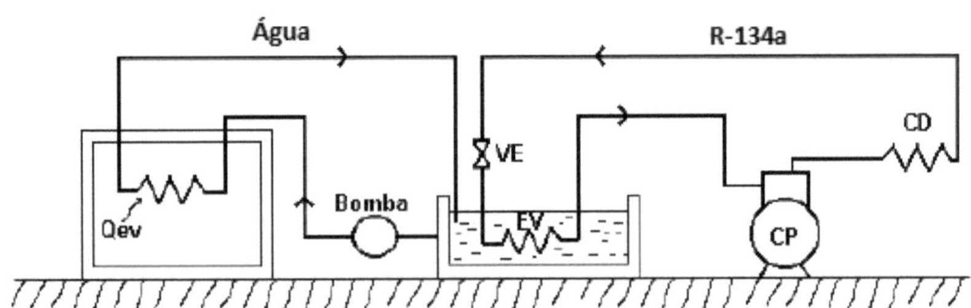

Figure 7 Indirect expansion system
Source: (SERAFIM MATOS, 2017)

Where:

Q_ev - Refrigeration capacity R-134a- Primary refrigerant EV- Primary refrigerant evaporator VE- Expansion valve

CD- Primary fluid condenser Water- Secondary refrigerant fluid

For large quantities of heat to be cooled, there are many applications for indirect systems such as commercial refrigeration, heat pumps, artificial ice rinks, larger dairies and slaughterhouses. In these cases, chillers are more economical and less complex than direct expansion systems (Handbook on indirect refrigeration).

For applications where a lower temperature level is desired, such as ice rinks, a component antifreeze, most commonly ethylene or propylene glycol, is mixed with the water and the term brine, in which salt has been used as antifreeze, is used to describe the secondary coolant (Stanford H.W.).

In some applications, cooling is essential, such as in computer centers, hospitals, laboratories, pharmaceutical or textile manufacturing, etc. Multiple chillers with at least one redundant chiller are often used. In this case, even if one chiller fails, 100% of the design cooling load can still be met (Stanford H.W.).

Long connecting pipes are common in air conditioning plants, artificial ice and ground source heat pumps with long collector pipes in the ground or in rock. Applications such as commercial refrigeration or supermarket refrigeration are increasingly moving towards indirect systems due to the need to minimize refrigerant charge and leakage (Handbook on indirect refrigeration).

The primary fluid can be condensed with water or air, depending on the project. Water condensing systems use cooling towers and are most commonly used when a higher thermal load is required, Figure 10.

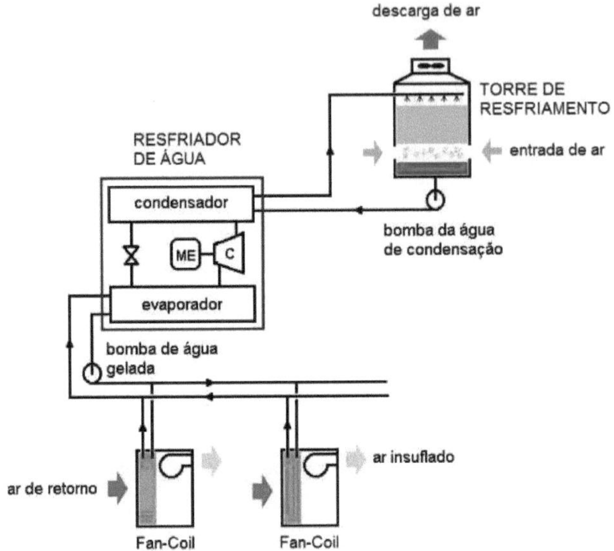

Figure 8 : Chiller with water condensation
Source: (CORRÊA, 2017)

3.6 THERMOACCUMULATION

Recent energy management policies have led to a constant improvement in energy sources. Thermoaccumulation is an effective way of making energy available, and is considered an energy efficiency alternative. This method of transferring energy does not reduce consumption, but rather makes it available from a lower tariff time to a higher tariff time, thereby reducing costs. (PUC- RIO, 2017).

This technology, which can lighten the load on the energy sector, has been explored for some time, but is still not widespread in Brazil and has few incentive policies. Thermoaccumulation is a way of storing energy for periods of more critical consumption. It can be stored in ice banks and chilled water tanks, and its greatest applications are in sectors that require large loads and use energy at peak times, which are interested in transferring the load from this period, when tariffs are higher, to another with a reduced tariff. (PUC-RIO, 2017)

There are different ways of storing energy, such as ice tanks, ice water and phase-changing substances like eutectic salts. They differ in the amount of energy stored per unit volume and in the temperatures of cold storage.

This water, stored in tanks, usually operates between 4°C and 6°C. The physical way of storing cold is through the sensible heat property of chilled water, so the volume that will be stored depends on the temperature difference between the feed water leaving the storage and the return water, the maximum difference being 11°C, with the possibility of this value reaching 17°C. The minimum volume of chilled water is 0.86 m^3/ kWh for 11°C (PUC-RIO, 2017).

As the water evaporates at a higher temperature, the energy needed to power the chiller circuit will also be lower, with the possibility of combining the chilled water tank with the fire tank.

3.7 ACCESSORIES

3.7.1 Expansion valve thermostatic

This valve, which can be called a superheat valve or thermal expansion valve, has a mechanism that regulates the flow of refrigerant to the evaporator according to the thermal load of the refrigerated environment (A. S. de CAMPOS & A. de SOUZA JR, 2009).

The more heat absorbed by the evaporator, the higher the temperature of the water vapor at its outlet. In the valve body there is a sensor bulb with a gas inside it so that when the temperature rises, the gas pressure increases, pushing the diaphragm, causing the shutter to open to increase the flow of refrigerant. In the case of a drop in temperature in the evaporator, the pressure in the sensor bulb decreases and the spring force in the diaphragm pushes the shutter to retain the flow of refrigerant. In both situations, the superheat specified by the spring pressure is maintained, as shown in Figure 11 and Figure 12 (CORRÊA, 2017).

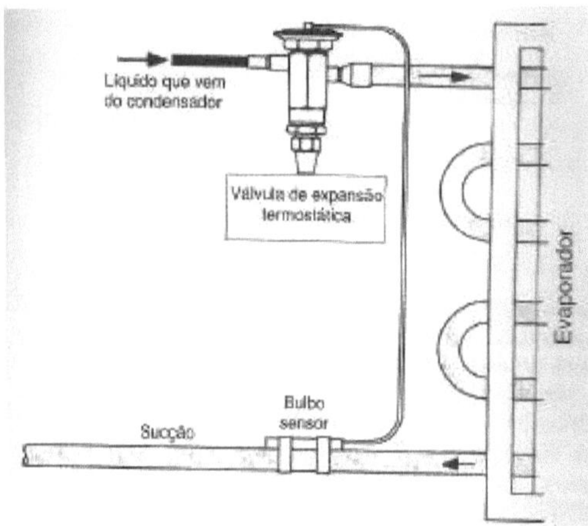

Figure 9: Schematic diagram of thermostatic expansion valve (TEV)
Source: (A. S. de CAMPOS & A. de SOUZA JR, 2009)

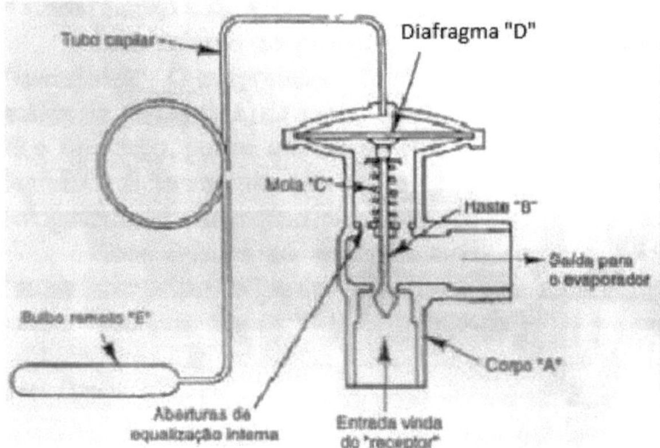

Figure 10: Schematic diagram of thermostatic expansion valve (TEV)
Source: (A. S. de CAMPOS & A. de SOUZA JR, 2009)

3.7.2 Filter dryer

Inside the refrigerant circuit piping there can be small impurities that can damage the components. Solid particles and water are some of these harmful elements. This moisture (water) freezes at 0°C and can obstruct the passage of refrigerant and form acids with corrosion potential. A cartridge made of a hygroscopic material - silica gel - and a binder (porous material) is inserted into a cylindrical metal housing. The cartridge may or may not be interchangeable and the filter is installed in the liquid line. Figure 13.

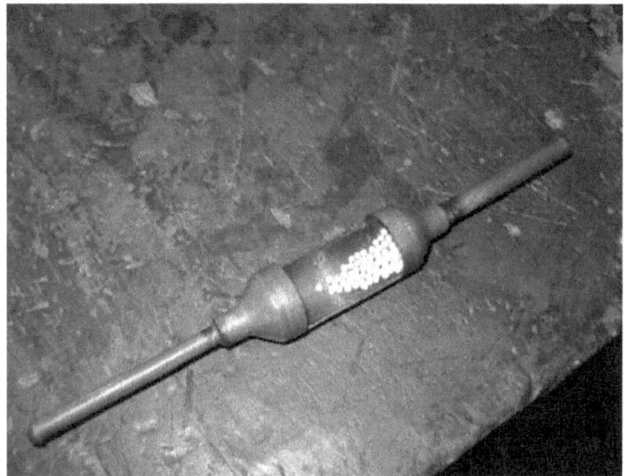

Figure 11: Filter drier
Source: (A. S. de CAMPOS & A. de SOUZA JR, 2009)

3.7.3 liquid display

In refrigeration, there is great concern about the appearance of humidity and refrigerant vapors in the evaporator, which may be due to an incomplete refrigerant charge in the installation. With the indicator it is possible to check the contamination of the refrigerant As seen in Figure 14 (CORRÊA, 2017).

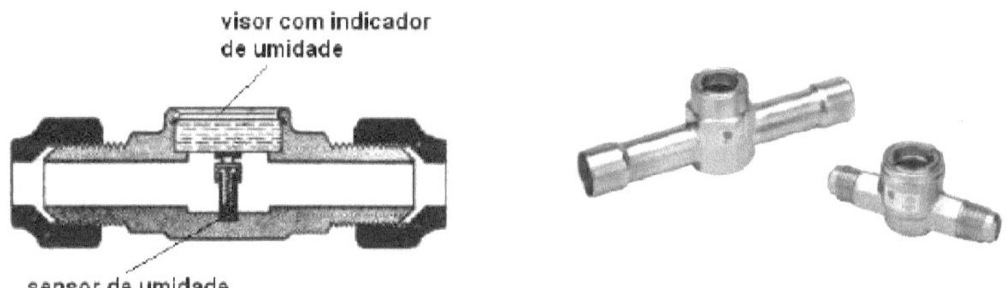

Figure 12: Liquid sight glass with humidity indicator
Source: (CORRÊA, Apostila de refrigeração e climatização, 2017)

3.7.4 Pressure switch electromechanical

To protect the system against any breakdown and automatic control, the pressure switch is very useful. When the pressure at the compressor inlet falls below a minimum permitted value, the low pressure switch switches off the compressor's electric motor. Similarly, if the pressure at the compressor outlet exceeds the maximum permitted limit, the high pressure switch switches off the compressor's electric motor. These two types can come together or separately or with regulation and without regulation, Figure 15, (A. S. de CAMPOS & A. de SOUZA JR, 2009)

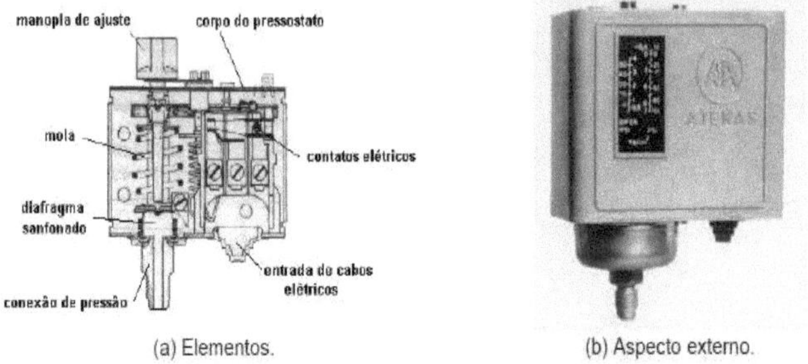

Figure 13: Pressure switch (a) Elements and (b) External appearance. **Source**: (CORRÊA, Apostila de refrigeração e climatização, 2017)

3.7.5 Thermostat

These are electromechanical devices used to control temperature, i.e. to maintain a pre-set average ambient temperature for refrigeration and air conditioning applications. As with the thermostatic expansion valve, the pressure switch has a sensor bulb containing gas which, when the temperature drops, the pressure of this gas also drops, disconnecting the electrical contact. When the temperature of the refrigerated environment rises, the gas pressure in the sensor bulb increases, pushing the diaphragm against the fixed contact, closing the circuit and powering the compressor Figure 16, (A. S. de CAMPOS & A. de SOUZA JR, 2009).

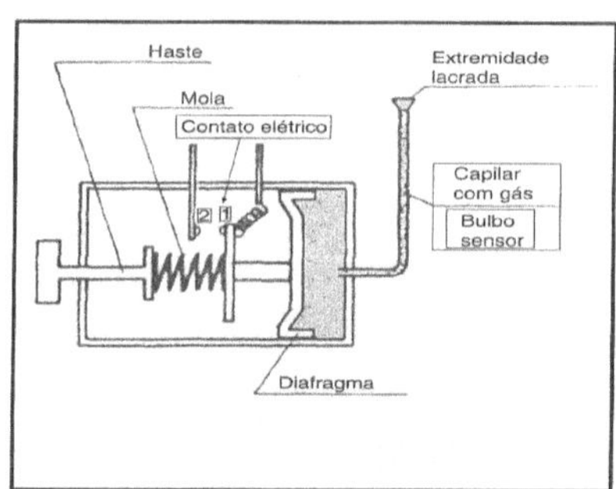

Figure 14: Schematic diagram of a thermomechanical thermostat
Source: (A. S. de CAMPOS & A. de SOUZA JR, 2009)

3.7.6 Separator from oil

The compressor is a mechanical component that needs lubricating oil. Therefore, a certain amount of oil is placed in its crankcase, but in such a way that there is no mixing of refrigerant and oil along the pipe, as the latter can deposit at low temperatures and create thermal resistance. To avoid this, an oil separator is placed just after the compressor outlet with the oil drained into the crankcase Figure 17, (CORRÊA, 2017).

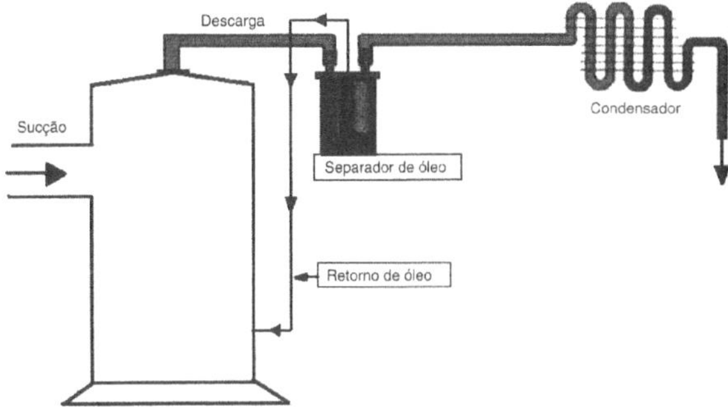

Figure 15: Oil separator
Source: (A. S. de CAMPOS & A. de SOUZA JR, 2009)

3.7.7 Valve from Service

Periodically or for an extra reason, the refrigeration machine undergoes maintenance such as vacuum services or gas refills. To carry out these activities, the service valves on the low and high pressure pipes are used. As the position of the valve varies, it releases the refrigerant for service or normal circulation Figure 18, (A.S. de CAMPOS & A. de SOUZA JR, 2009).

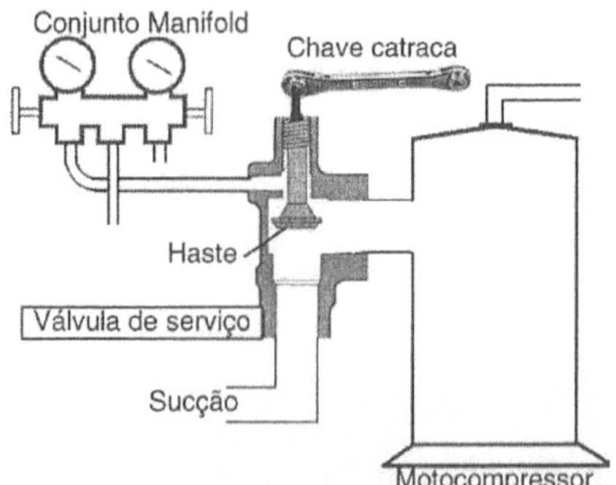

Figure 16: Service valve
Source: (A. S. de CAMPOS & A. de SOUZA JR, 2009)

3.7.8 Accumulator from suction

In order to ensure that no liquid passes through to the compressor, a suction accumulator or liquid separator is used at the condenser outlet for complete evaporation. They are most commonly used for freezers, chambers and refrigerated counters that require low evaporation pressures Figure 19, (A. S. de CAMPOS & A. de SOUZA JR, 2009).

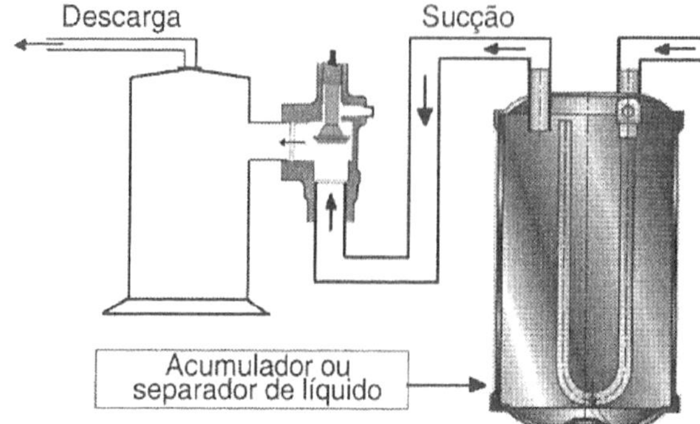

Figure 17: Suction accumulator or liquid separator
Source: (A. S. de CAMPOS & A. de SOUZA JR, 2009).

3.7.9 Tank of Liquid

This very important component is installed just after the condenser and can store all the liquid in the system. For two reasons, either for general maintenance or if the condenser fails for some reason, the tank will provide a supply of refrigerant for the liquid line, Figure 20, (A. S. de CAMPOS & A. de SOUZA JR, 2009).

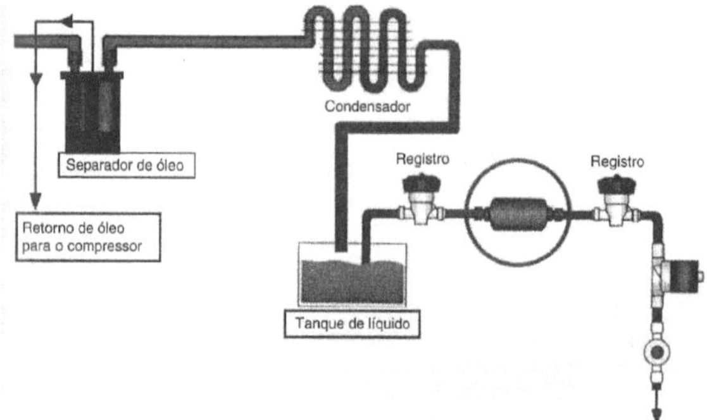

Figure 18: Liquid tank
Source: (A. S. de CAMPOS & A. de SOUZA JR, 2009)

3.7.10 Flexible tube

To dissipate compressor vibrations, a flexible pipe is used in the installation, which can be in the low or high pressure line

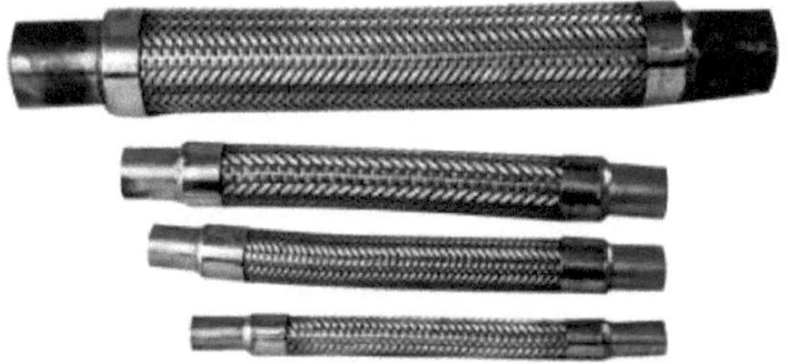

Figure 19: Flexible tube
Source: (Getrat, 2016)

3.7.11 Evaporators

Component in which the refrigerant absorbs latent heat and changes from liquid to gaseous phase, i.e. the part that exchanges heat with the air to be cooled. During vaporization, the pressure drop between the inlet and outlet of the evaporator must not be excessive and there must be sufficient area and length to evaporate the entire liquid-vapor mixture entering the equipment. (A. S. de CAMPOS & A. de SOUZA JR, 2009)

The most common are the dry type, in which the refrigerant enters as a mixture of liquid and vapor and exits entirely as vapor (CORRÊA, Apostila de refrigeração e climatização, 2017). They can be smooth tube, finned tube, plate and circular tube. In circular tubes, there is a loss of pressure along the pipe, causing it to drop dramatically at the end of the evaporator Figure 22.

Figure 20: Tube evaporator
Source: (Getel equipment, 2017)

3.7.12 Capacitors

These are heat exchangers in which there is a phase change from vapor to liquid through the loss of heat from the refrigerant. Condensation can be air- or water-cooled and pressure drop levels, physical space, location, costs and environmental aspects must all be taken into account when choosing an evaporator. For air-cooled condensers, there are capacity ranges from 3.5KW to 351.7 KW with natural or forced transmission, Figure 23.

Figure 21: Air-cooled condenser
Source: (SERAFIM MATOS, 2017)

3.7.13 Fans

Fans are flow machines that maintain a constant flow of air by means of a pressure difference. In the housing of this component, dynamic pressure is converted into static pressure, and the ratio between them is generally no more than 1.07. They can be centrifugal or axial, in which case the air enters the fan in the same direction as the motor shaft and exits at 90° and in axial fans the air at the inlet and outlet is parallel to the motor shaft, Figure 24, (CORRÊA, Apostila de refrigeração e climatização, 2017).

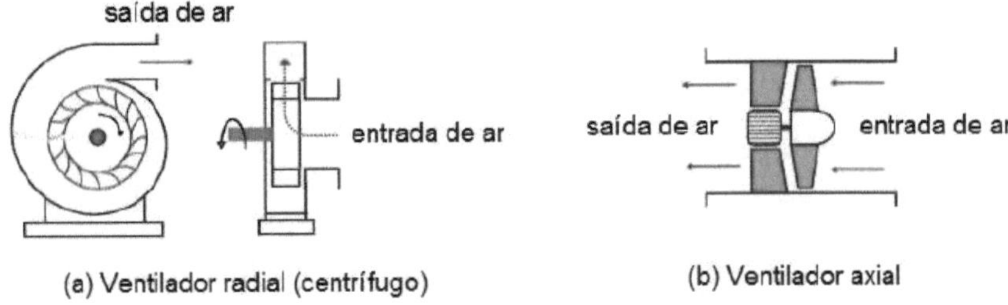

Figure 22: Direction and direction of air flow in fan types.
Source- (CORRÊA, Apostila de refrigeração e climatização, 2017).

3.7.14 Compressors

The power input source in the refrigeration cycle, responsible for moving heat from a lower temperature source to a higher temperature source. They can be reciprocating, centrifugal or rotary. In reciprocating units, the gas is compressed by reducing its volume and they have different constructions, such as open, semi-hermetic and closed. In open units, the piston shaft is driven by the outer shaft of the engine, Figure 25, (SERAFIM MATOS, 2017).

Figure 23: Open reciprocating compressor
Source: (TR refrigeração e Climatização, 2017)

3.8 ARDUIN

Arduino is a microcontroller board in an electronic prototyping platform of Italian origin developed by Massimo Banzi and David Cuartielles in 2005, which among its proposals is to facilitate the elaboration and development of projects of the most varied degrees of complexity. With this instrument, various actions can be carried out, such as controlling motors, temperature sensors, movement, LEDs, robotics and other electronic components (Eletrogate, 2017).

There are several models of hardware boards on the market, as there are boards designed for each project, which vary in price in Brazil from around R$ 60.00 to R$ 80.00, so that the larger boards can control more electronic devices. Its internal workings are similar to those of a computer with reduced features and the following components (Eletrogate, 2017);

1. Microprocessor (which is responsible for calculations and decision-making);
2. Ram memory (this stores data and volatile instructions);
3. Flash memory (Used to store software, non-volatile);
4. Timers;
5. Accountants;
6. Clock, etc.

The Uno model has some specifications such as: 14 digital input/output pins, 6 of which can be used as PWM (analog) outputs, a USB connection, 6 analog inputs, a 16 MHZ ceramic resonator, an ICSP header,a power connector that can be connected either to a computer or to an AC to DC power source or battery (Arduíno mais, 2014).

Power is supplied via USB with a DC voltage of 5v, which is supplied when connected to the computer. Higher voltages of 7.5 and 12 V can be used with a p4 plug or directly on the board using the Vin pins and ground pins. (Arduino mais, 2014)

According to (Álvaro Justin, 2017, p. 15) "Arduino has two types of input ports: analog and digital. In addition, the digital ports also serve as output ports, working with two basic types of common digital output and PWM output that can be used to simulate an analog output, among other things." Figure 26 shows the image of a digital Arduino.

Figure 24: Digital ports 13 to 0
Source (Álvaro Justen, 2017)

We also have analog ports, which differ from digital ports in that they are only for data input with a power supply in the range of 0 to 5 v , Figure 27

Figure 25: Analog ports one to 5
Source (Álvaro Justen, 2017)

3.9 PSYCHROMETRIC PARAMETERS

The composition of the air we breathe is made up of various gases, water vapors and contaminants such as dust, pollens, polluting gases, etc. When we remove all the water vapor and contaminants, we have dry air, which is an ideal model since in reality it is not possible to remove all the water vapor and contaminants. This ideal composition is almost constant, with small variations in the composition of some air components depending on geographical location, time of day and altitude. Table 2 below shows the composition of dry air in the atmosphere (Jorge and Corrêa).

Table 2: Composition of dry air in the atmosphere

COMPONENTE	VOLUME (%)
1. Nitrogênio	78,084
2. Oxigênio	20,9476
3. Argônio	0,934
2. Dióxido de carbono	0,0314
5. Neônio	0,001818
6. Hélio	0,000524
7. Metano	0,00015
8. Dióxido de enxofre	0 até 0,0001
9. Hidrogênio	0,00005
10. Kriptônio, xenônio e ozônio	0,0002

Source: Climatização e refrigeração Jorge E. Corrêa

Humidity is present in small or large proportions. A mixture of dry air and water vapor is called humid air, in which the amount of water vapor varies from zero to a maximum, depending on the temperature and pressure of the mixture. When the air is excessively humid, it is said to be saturated.

The air temperature in air conditioning applications ranges from around -10 °C to 50°C. In this range, dry air can be treated as an ideal gas with a constant cp equal to 1,005 kj/kg.K [0.240 BTU/lbm.R] with negligible error (below 2.0 %). Considering 0°C as the reference temperature, the enthalpy and enthalpy change of dry air can be determined by: (ÇENGEL,2013, p.732)

$$h\text{-}□□□e□o=□\text{-}□=(1.005kj/kg.{}^{\circ}C)T \qquad (kj/kg) \qquad (3\text{-}14)$$

$$\Delta h\_arseco=c\_p\,\Delta□=(1.005kj/kg.{}^{\circ}C)\Delta T \qquad (kj/kg) \qquad (3\text{-}15)$$

3.9.1 Relative humidity air

"Relative humidity is the approximate relationship between the masses of water vapor present in a volume and the mass of vapor that would saturate that volume at the same temperature and total pressure." (PENA, 2012, p. 10)

$$UR = m\_vapH20/m\_vapH20\text{max} \qquad (3\text{-}16)$$

Relative humidity is most commonly measured using psychrometers, which basically consist of two thermometers, one dry bulb and one wet bulb. According to Çengel (2013,p.743) "most people prefer a relative humidity of 40 to 60%."

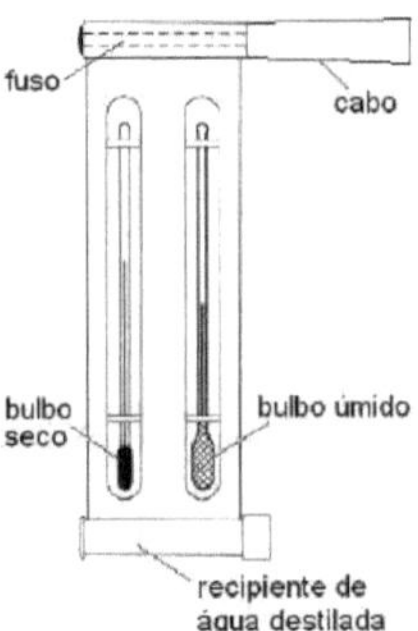

(a) (b)

Figure 26: Psychrometers: a) digital and b) rotating
Source: (Vectus, 2017)

3.9.2 Dry bulb temperature of air

It is generally always higher than the wet bulb temperature and is the temperature indicated for the air-vapor mixture by a common thermometer. This temperature is the same for both elements of the mixture, i.e. steam and air. (PENA, 2012)

3.9.3 Wet bulb temperature of air

"This temperature is obtained by a thermometer where its bulb wrapped in wet gauze is exposed to a current of air until the equilibrium temperature of the air-vapor/bulb mixture is obtained and the temperature stops dropping. This temperature will be lower than the Dry Bulb temperature" (PENA, 2012, p. 17).

3.9.4 Enthalpy

A thermodynamic property that measures energy in the form of heat, whether the fluid is at rest or moving. In general, enthalpy differences are used. This difference corresponds to the amount of heat exchanged by the air (air-vapor mixture) between two positions. (PENA, 2012)

3.9.5 Chart psychrometric

It is a graph that allows you to trace various psychrometric processes and determine the states and properties of humid air. It is an essential ally for refrigeration and air-conditioning designers to quickly analyze and quantify the various processes that humid air goes through as it flows through equipment, accessories and devices. It is based on the fact that the thermodynamic state of humid air is determined by three independent properties (CORRÊA, Apostila de refrigeração e climatização, 2017, p. 81).

It can be read as a two-axis graph when one of the three properties is kept constant. This constructs the isolines of the diagram and any point on the graph defines a state of the mixture, i.e. humid air. Generally, the pressure of the mixture is chosen as the property to be kept constant, since in most psychrometric processes its variation is negligible. (CORRÊA, Apostila de refrigeração e climatização, 2017), The quantities represented in this diagram are as follows:

- Dry bulb temperature - BS- $^{o.}$ C or r
- Wet bulb temperature - BU- $^{o.}$ C or r
- Relative humidity - (Uk)- %
- Absolute humidity (W) - lb vapor/lb dry air. or grains vapor/lb dry air or grams (g) vapor/ kg dry air
- Specific Volume - (v) - m3/kg air or ft 3of air/ lb air - (the specific volume is the

inverse of the specific mass).

- Enthalpy- (h. or i)- BTU/ lb air or kcal/kg air

The following figures show the reading scheme and a typical psychrometric chart, respectively, Figure 29 and 30.

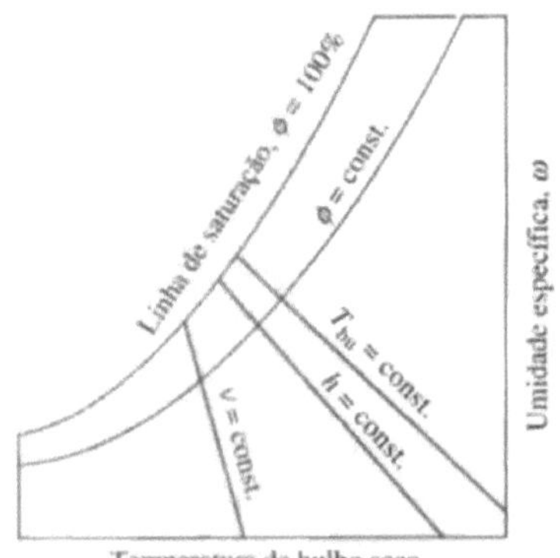

Figure 27: Schematic representation of a psychrometric diagram
Source: (ÇENGEL,2013)

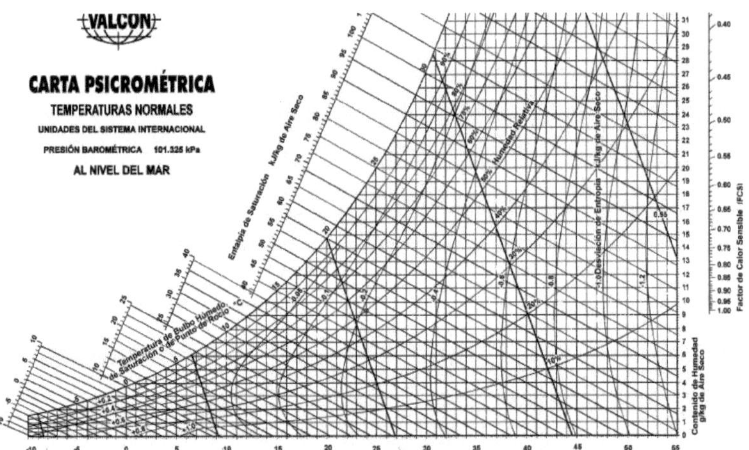

Figure 28: Psychrometric chart
Source: (Psicrometría, 2013)

4 MATERIALS AND METHODS

The bibliographic database for this academic work was obtained from engineering books in university libraries, technical manuals on air conditioning and refrigeration, related articles and handouts available on the internet. The experiment was carried out in the refrigeration laboratory at the State University of Maranhão-UEMA.

In order to collect data from the machine, a number of procedures and applications of tools and materials were carried out. Before starting up the machine, the system was vacuumed in order to remove possible air and water that could cause chemical attacks on the copper pipes, clog holes with unnecessary pressure rise and other damage. This process is carried out using a vacuum pump and manifold assembly similar to those shown in figure 31 below.

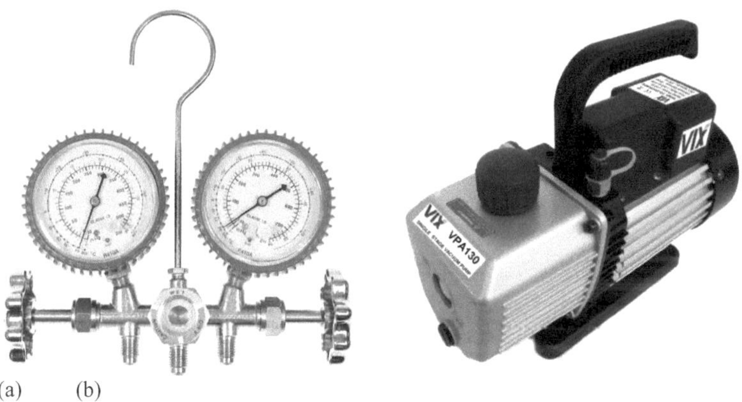

(a) (b)

Figure 29: a) Manifold and b) Vacuum pump
Source: (Frio peças, 2015)

Vacuuming was carried out with the refrigeration machine switched off. Initially, we put the service hose (yellow) from the manifold pressure gauge on the vacuum pump and the other two hoses, blue and red, on the low and high service valves respectively on the compressor. We then briefly release these valves in order to carry out the service. With the manifold registers closed, we start the pump to initially make the vacuum from the service hose and a few seconds later the high and low pressure registers are released simultaneously.

The manifold's low-pressure clock takes a vacuum reading on a scale of up to 30 microns of mercury (30 in Hg), with the factory recommending that a good vacuum is one between 250 and 500 microns of Hg. To take this reading, an electronic vacuum gauge must be used. At the end of the evacuation, we close the manifold registers and check the movement of the pointer on the low pressure clock. Ideally, the pointer should register 30 in Hg, but the pointer dropped to 20 in Hg, indicating that there is a gas leak in the system Figure 32.

Figure 30: Evacuation
Source: (Author)

Once the vacuum is made, the next step is to recharge the system's refrigerant gas. We used an R134a gas cylinder and a manifold to carry out the recharging process. With the service hose uncoupled, we opened the cylinder damper slightly to release any air that might be present. With the three hoses in place, the manifold valves and the cylinder damper, which is in the vertical position characterizing a steam phase recharge, are opened until the pressures stabilize. The machine is then switched on so that the compressor can complete the gas recharge. The low pressure manifold clock registered 85 psi during the recharge. It is worth noting that the manifold hoses do not have the sealing ring in good condition, which can result in an incomplete vacuum Figure 32.

Figure 31: R134a refrigerant gas recharge
Source: (Author)

Once the system had been vacuumed and recharged with gas, the machine's pressures and temperatures were collected. In order to record the inlet and outlet temperatures of the condenser and evaporator more accurately, an Arduino Uno microcontroller with wired sensors and a Protobord board with a Microsoft Excel interface was used. Programming was carried out in Excel's VBA (Visual Basic for Application) environment to make the microcontroller act as a thermometer and record the temperatures every 10 seconds.

The experiment took approximately 53 minutes. All the temperatures collected by the sensors were automatically sent to a database in an Excel spreadsheet Figure 34.

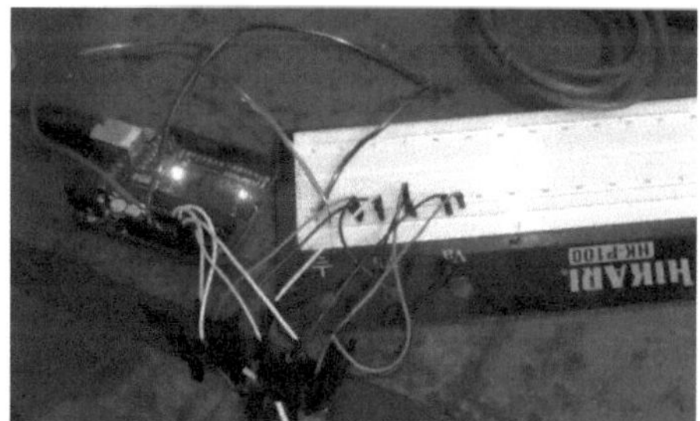

Figure 32: Arduino kit
Source: (Author)

Below is part of the Arduino programming for the temperature sensor in VBA, Figure 35.

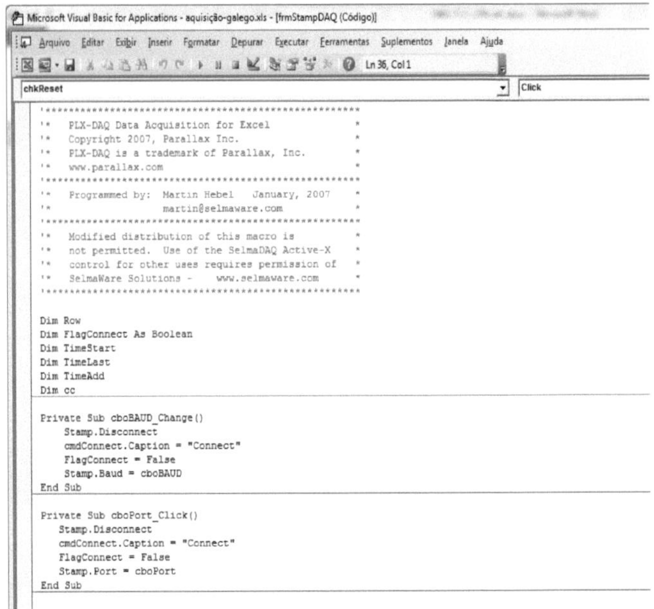

Figure 33: VBA programming environment
Source: (Author)

5 RESULTS AND DISCUSSIONS

With the data sent to EXCEL via the microcontroller, we have a distribution of temperatures over time in the components analyzed, the condenser and the evaporator. Over the 53 minutes starting at 12:16:16 and ending at 13:05:50 during which the experiment was carried out, the temperature dropped by approximately 3 degrees, from 28.50 °C to 25.2 6°C. As the time interval for recording data in the spreadsheet is 10 seconds, it becomes several bands with practically identical data, so the parts where there are different temperature values have been selected, as seen in Table 3.

Table 3: Evaporator temperature table

Evaporator		
Time	Inlet temperature	Output temperature
12:16:16	28,50	29,6
12:16:26	28,50	29,7
12:16:36	28,50	29,7
12:16:46	28,50	29,7
12:36:48	27,59	29,05
12:36:58	27,41	28,96
12:37:08	27,32	28,87
12:37:18	27,14	28,87
12:40:58	26,51	27,96
12:41:08	26,33	27,86
12:41:18	26,24	27,77
12:41:28	26,06	27,68
13:05:20	25,35	26,42
13:05:30	25,35	26,42
13:05:40	25,35	26,42
13:05:50	25,26	26,42

Source: (Author, 2017)

This temperature distribution can be better observed in the graph below, Figure 36.

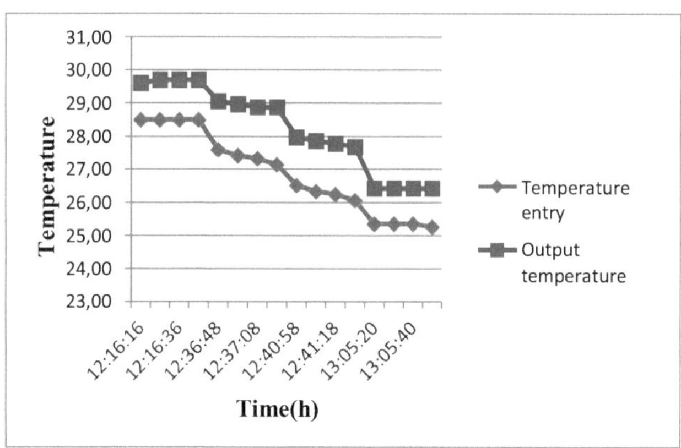

Figure 34: Temperature evolution in the evaporator
Source: (Author, 2017)

We can see that there is a decrease in temperatures in the condenser until the end of the experiment, but the difference in temperature at the inlet and outlet is very small, characterizing a very large inefficiency of the machine, since the evaporation temperature of the refrigerant leaving the thermostatic expansion valve is high. The same analysis was carried out for the condenser, obtaining the following data, Table 4:

Table 4: Condenser temperature table

Condenser		
Time	Inlet temperature	Output temperature
12:16:16	39,42	34,0510047
12:16:26	39,32	34,0520056
12:16:36	39,32	34,0530063
12:16:46	39,42	34,0540064
12:36:48	40,06	34,25124033
12:36:58	40,17	34,35125033
12:37:08	40,17	34,15126034
12:37:18	40,17	34,25127034
12:40:58	40,06	34,1514904
12:41:08	40,06	34,1515004
12:41:18	40,06	34,1515104
12:41:28	40,06	34,15152041
13:05:20	35,03	31,6629508
13:05:30	35,03	31,6629608
13:05:40	34,93	31,6629708
13:05:50	35,03	31,5629808

Source: (Author)

The temperature evolution can best be seen in the following graph, Figure 37.

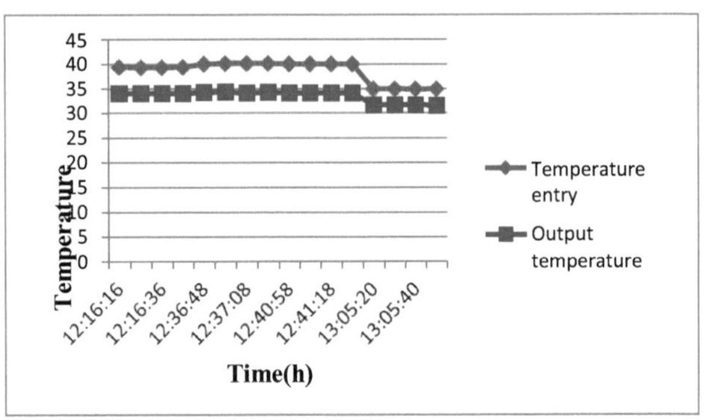

Figure 35: Temperature evolution in the condenser
Source: (Author)

We can see that the temperature of the refrigerant drops as it leaves the evaporator and remains practically constant over time. However, the temperature differences are low, which characterizes the machine's inefficiency. When the system stabilizes, with an approximate temperature of 25.26 °C in the evaporator, the low and high pressures remain at 5 psi and 75 psi respectively, whereas ideally these pressures should be around 20 psi and 200 psi, thus indicating anomalies in the system. The following figure shows the manifold pressure gauge clocks with the low pressure (blue) and high pressure (red) when the system was stabilized, Figure 38.

Figure 36 Pressure recording when the system stabilizes
Source: (Author, 2017)

6 CONCLUSIONS AND SUGGESTIONS FOR FUTURE WORK

A refrigeration simulator is of great importance for consolidating the knowledge acquired in the classroom on a Mechanical Engineering course, as it allows you to see a refrigeration circuit in practice with its components in operation, as well as being able to carry out various experiments and studies.

Through the experiments carried out in this work, we found that the laboratory's direct expansion refrigeration machine, even after the gas evacuation and recharging procedures, is operating at temperatures that are unsuitable for good heat absorption from the environment to be cooled, due to supposed problems with refrigerant leaks in the pipes or connections.

It could be suggested that after the machine has been properly maintained, new pressure gauges and flow measurement instruments should be used to increase data collection in order to calculate the machine's cooling capacity, pressure loss, absorbed and rejected heat, as well as interconnecting it to the chilled water circuit, the Chiller that is present in the refrigeration laboratory at UEMA.

REFERENCES

Arduino more (July 23, 2014). Accessed on June 17, 2017, available at Details about arduino uno: http://arduinomais.blogspot.com.br/2014/06/detalhes-sobre-o-arduino-uno.html

Stanford, H. (n.d.). HVAC Water Chillers and Cooling Towers: Fundamentals, Application and Operation.

Stanford, H. W. (n.d.). HVAC Water Chillers. In: *Fundamentals, Application and Operation.*

A. S. de CAMPOS, R., & A. de SOUZA JR, A. A. (2009). *Refrigeration and air conditioning workbook.*

Álvaro Justen (2017). Arduino course.

Cengel, Y. (n.d.). Thermodynamics. In: Y. Cengel, *Thermodynamics(2013)* . (S.D.). Air conditioning and refrigeration.

CORRÊA, P. D. (2017). Refrigeration and air conditioning handbook.

Eletrogate, c. e. (2017). *Basic Arduino handout v1.o*. Source: Eletrogate, electronic components.

Handbook on indirect refrigeration (n.d.).

PENA, E. S. (2012). Air conditioning and refrigeration systems. PUC-RIO (n.d.).

SERAFIM MATOS, D. (n.d.). Refrigeration.

Refrigeration Systems: Direct Expansion vs. Indirect Expansion (2016). Source: Diamont precisão em climatização.

(n.d.). Air conditioning and refrigeration. In: W. F. Stoecker, & Jerold W. Jones, *Air conditioning and refrigeration* (pp. 1-3).

(2009). Chiller System Design and Control. In: Susanna Hanson, Mick Schwedler, & Beth Bakkum, *Chiller System Design and Control* (pp. 19-20).

Printed by Books on Demand GmbH, Norderstedt / Germany